CW01151177

The Rise of AI Gadgets: Transforming Everyday Life

Leonard Crimson

Published by RWG Publishing, 2024.

While every precaution has been taken in the preparation of this book, the publisher assumes no responsibility for errors or omissions, or for damages resulting from the use of the information contained herein.

THE RISE OF AI GADGETS: TRANSFORMING EVERYDAY LIFE

First edition. May 28, 2024.

Copyright © 2024 Leonard Crimson.

Written by Leonard Crimson.

Table of Contents

Introduction to AI Gadgets 1
Neural Headphones: The Future of Audio Technology 5
Smart Rings: Wearable AI at Your Fingertips 9
AI Assistants: The Personalized Virtual Helpers 13
AI Gadgets in Health and Wellness 17
AI Gadgets in Home Automation 21
Ethical and Privacy Considerations in AI Gadgets 25
Future Trends and Innovations in AI Gadgets 29

Introduction to AI Gadgets

As far as AI gadgets go, it is not just about improving products or providing user assistance. The true value lies in the accompanying intelligence and learning capabilities of these gadgets, based on the AI. They have a much broader role to play in our lives. As shown through our writings, AI gadgets offer both business and consumers extended choices to communicate, obtain new and richer data sources, make faster and better decisions, and eventually generate more value. These contributions of AI gadgets, particularly with the spread of their increasing applications and interfacing, reflect a growing realism rather than hype in the provision of products and services. Model predictions estimate that AI could contribute about $16 trillion globally by 2030, more than the current output of China and India combined. The launching and acquiring of AI gadgets demonstrate well the pace at which enterprises seek to exploit the potential created by AI.

As technology improves, so does the sophistication of smart gadgets. AI gadgets, or those with artificial intelligence embedded in them, are accelerating the future belonging to smart technology. This advancement is proven to be practical and accessible, in comparison to the futuristic image most often associated with science fiction literature and movies. Unbeknown to many, we are already living this reality. Technological advancements in areas such as industrial robotics and drones, among others, which are already contributing to global productivity, often escape notice. However, from Siri to Alexa, smart gadgets that employ artificial intelligence to assist, inform, or entertain, are now a part of our daily existence, regardless of us embracing them wholeheartedly or not.

1.1. Definition and Scope of AI Gadgets

In terms of the definition, as disclosed in its naming, AI gadgets generally refer to popularized products embedded with artificial intelligence techniques, but unlike traditional AI hardware or software that add machine intelligence to products which are not originally intelligent by their natures, AI gadgets possess intelligent characteristics but are not intelligent itself. In specific technology, AI gadgets widely employ multiple AI techniques, but mainly focus on perception and understandable optimizable decision-making. They leverage embodied intelligence to support people in everyday tasks, go further in helping users relieve pressures from routine issues, and promote the social acceptance of AI technologies.

AI gadgets form an emerging area in AI research and practice, with the aim of integrating AI technology with existing popular gadget products in daily usage. This emergent area has just started to be recognized and analyzed. Lacking a systematic study, AI gadgets are various in form factors and AI techniques, due to their fuzzy definition, uncertain targets, multiple design purposes, and diverse technical basis. This chapter thus aims to propose a consistent and integral definition of AI gadgets, turning discussions based on the proposed definition to elucidate the scope and boundary of AI gadgets, to make a systematic discussion of AI gadgets at a more fundamental level. It is also important in practice, e.g., from identifying potential product directions, to analyzing exploitation chances for various techniques.

1.2. Historical Background and Evolution

Consider the development of mankind as a process that has consisted of different models of "intelligent systems." Today, we recognize four main models: the instinctive model, the memorizing model, the programmed model, and the learning model. The modern microprocessor is much more related to the programmed model than to the others, which is completely different from all the learning of the human mind. The learning systems are relatively late technologies. The emerging science of artificial intelligence seeks to create machines capable of engaging in intelligent activities. Such a machine would produce words, concepts, images, thoughts, and other products of human intelligence. AI is an interdisciplinary science with multiple approaches, but advancements in AI have been rapid and remarkable over the last 50 years. If its present rate of progress continues, we will soon be able to build AI machines whose intelligence rivals that of any human being.

A brief historical perspective on man's quest for intelligence amplification aids in our understanding of the developments that have led to AI gadgets. It has been stated that a key driver of the evolution of civilization is the development of technology that will facilitate man's work. From the use of simple tools that enabled lifting and carrying to today's space-age technology, man's aim has always been to work better. Technology has been employed in almost every aspect of life - from simple needs such as food, shelter, and clothing to complex ones such as business, health, and entertainment. Over time, technology has evolved significantly from tools that ease muscle usage to tools that ease decision-making processes. The introduction of computers pumps up the speed of analytical processes.

Neural Headphones: The Future of Audio Technology

The technical details of this solution are still a mystery, but what is known is that it will come with the Google Pixel OS, allowing the user to freely control their playlist. This will be carried out through voice control - the system should be based on the voice commands given to Human, launched on the Amazon Alexa platform. The headphones will feature several additional smart features of a hearing aid, including active noise exclusion.

More functions, less devices - this is the idea that motivated the team of Human, the robotic assistant overcoming blindness and memory problems launched in 2019. And now the company is on the verge of launching Neural Headphones, the headphones capable of doing the job of hearing aids, while, at the same time, streaming music from any digital platform. The central idea of this project is to match different functions in one single device. The headphones should not only help people with hearing loss, who have no interest in losing the benefits of music, to hear better, but also make life less complicated. Therefore, instead of being obliged to purchase a hearing aid and yet another device to stream music, Neural Headphones will offer both at the expense of a single device.

2.1. How Neural Headphones Work

From the consumer's perspective, these neural earphones make AI assistants much more accessible and, as a result, much more likely to be used regularly. Voice-activated home assistants like Google Home Mini, Amazon Alexa, and Apple HomePod might seem cool, but to use them means buying at least one new device and finding shelf space for something that always needs to be listening. Ear-assisting devices are always listening too, but they are designed to only listen for a single code word, like Alexa or Hey Google. What's the big deal, you ask?

It's whatever when that code word tells these ear-assisted AI gadgets to start doing amazing things quietly (even remotely) and privately. In other words, they work as great, natural, and out of the way voice interfaces to the ever-growing abundance of smart products and digital services available to users these days.

To get a better sense of how exactly these headphones work, it's crucial to have a bit of knowledge about AI algorithms which power up tools like Alexa, Bixby, and Siri. Designed engineers specialize in developing advanced machine learning models which allow AI devices to process and understand various types of real-world data, including human speech. This advance alone is similar to comparing a flashlight to an artificial light source like an LED. Although both help us 'see,' one can work in the sun, underwater or on the moon. The other can be 'smart.' Its light can change color, turn on and off in response to simple commands, brighten or dim, and much more.

2.2. Key Features and Benefits

More often than not, consumers save time and effort, are more efficient, organized and accurate, inspired to learn, become healthier or reach fitness goals, access specific learning programs or become more creative. Primarily, AI gadgets help consumers improve their lives using the latest innovations, adjust to personalized data services and proactively inform consumers in specific areas. After familiarizing the reader with the key features and benefits of AI gadgets, several power and propelling forces are overviewed.

AI gadgets have different features and the benefits they bring to consumers vary as well. Not so long ago, IoT was seen as a technological evolution of digitization, resulting in smart gadgets that could collect, process, and exchange information with each other or the use of mobile applications that made smartphone users' lives more comfortable. In the last decade, various types of electronic gadgets with integrated AI functionalities have been introduced in the marketplace, with a series of specific features that have not been seen before, such as providing personalized data services, proactively informing consumers in specific areas or controlling various environments or objects. They are constantly learning and adapting to ensure they cater to the personalized tastes and needs of the user.

Smart Rings: Wearable AI at Your Fingertips

The hands-free aspect of smart glasses during unobtrusive activity, such as driving, cycling, and running, translates into its hands-free convenience in daily life. Fewer devices are closely related to hands, such as rings. Wearable electronic gadgets, especially smart rings, are increasingly becoming a part of the AI field. While embedding wireless capabilities, biometric authentication, and health monitoring functions, rings still maintain their fashionable exterior and chic functions. For example, a ring can monitor vital signs by sensing the pulse under the skin, which provides doctors with a tool to record resting heart rate continuously. Wearable AI technology is innovative, but space is rather limited. Therefore, studies regarding their comfort and functionality are meaningful and interesting. AI gadgets, including smart rings, show a clear trend of unobtrusiveness in embedded capabilities.

3.1. Functionality and Applications

For instance, you can find smart displays showing weather info or pictures like a digital photo frame, smart speakers with a bigger screen playing music videos, smart thermostats with built-in speakers, robot vacuum with a camera for security, LED lights that can play music, and so on. Also, most gadgets can recognize the human voice and respond through a natural language processing speech assistant. So, AI gadgets can be considered easy to use by anyone. These speech assistants, along with their AI, are powering up the acceleration of smart gadgets. Imagine what a smart fridge will be able to do in a few years when embedded with these technologies.

While not all AI gadgets are created equal, most brands offer useful functionality through their gadgets. Some gadgets can even put paid to several other gadgets in one go, as they can replace more than one of your smart devices. It is nothing new that you can watch TV, play music from a smart speaker, get weather info from a smart display, and control some home automation systems with these gadgets. The point is, these "smart life" common functionalities are in great demand because they are incorporated into just one product.

3.2. Design and User Experience

The AI tools and platforms learn from humans and continue the human legacy more quickly, freeing them to develop unimagined creativity. The AI system Savita has developed, AI Canvas, as an assistant to quickly generate and maintain the color palette, fabric print, and pattern library industries and thus provide the much-needed cost-effective assistance in meeting market demand. Moreover, according to the company, the platform respects cultures and makes traditionally challenging objectives (e.g., unisex dress) easy, affordable, and desired to see efficiently met.

Utilizing its own proprietary AI software, which uses algorithms that expose the dimensions and quantities of relationships and patterns in nature to generate color palettes, shapes, and fabric prints based on all historical representations of any specific matter, Savita develops what they call AI-Designed Mood Boards or HMI/KPIs. These, in essence, visually communicate the feelings, concepts, and criteria behind each of their AI-designed or human-designed brand identities, providing insights into the effectiveness of their unique relationship.

An impressive demonstration of how AI can significantly improve the design process took place recently at the Human 2020 Virtual Summit in the session "Getting Creative: How AI and Humans Are Partnering to Design a More Beautiful World," which was described as an "exploration across the divide between machine and human intelligence" and was organized jointly by the Massachusetts Design Art and Technology Institute (i-DAT) and Savita, a company that provides "the next generation of 3D design tools."

The rise of wearables and the advent of AI devices that are no longer disproportionately large, unattractive, and overly technical is essentially due to better-quality designs that have become far more affordable. The marriage of AI and design, called AI-assisted design, is still in its early phases, although AI is increasingly being employed not only in improving the speed and efficiency of design but also in creating designs that could not be done by normal designers, according to Jeremy Pava of the MIT Center for Bits and Atoms.

AI Assistants: The Personalized Virtual Helpers

The product becomes their personalized AI. Like a friend that knows a person's preferences as well as planning, the AI personalizes the product based on the users' image recognition and data inputs. They recognize familiar faces, bewilderment, family members, and guests, and will behave accordingly. They keep track of daily activity, make the product more personalized and flexible to meet the needs of their trusted user, and even recommend appropriate settings or actions. In addition to "performing their duties," the AI personalizes the product they are embedded into the step-by-step process towards other AI voice services. Users are also able to use endless third-party voice applications specially designed for the product to get the latest events, check on the latest news, power, ask questions, and others. Users can also link it with their smart health devices. This AI approach embeds into sleep tracking, analyzing, and making insightful suggestions.

Reliable, patient, and always willing to provide help. This is what describes what consumers find helpful in AI voice assistants. AI technology has become an essential tool to create a personalized, flexible, and enriched lifestyle. With AI voice assistants, the product provides more than just simple tasks that respond to questions asked by the user, but to map out a user's task by day along with his user's data trend. With this AI tool always close by, either at home or on the go, the product lets its customers interact with it using only their voices. Customers no longer have to type messages or press buttons to deploy how the product can help them - AI techniques make the products more than easy and fun to use, in which users can interact using an AI voice assistant in the same way that they interact with friends.

4.1. Popular AI Assistants in the Market

Apple's AI assistant, Siri, is one of the most widely recognized AI-based personal voice assistant software and has been joined by others in the form of Amazon's Alexa, Microsoft's Cortana, and Google Assistant and third-party AI assistants like Bixby for Samsung mobile devices and Duer for Baidu. All such assistants leverage a modern AI innovation, i.e., deep learning for specific tasks of text-to-speech and speech recognition supported by increasingly powerful GPU. Siri has been designed for the iOS platform and uses many areas of AI, including natural language processing and computer vision, to accomplish the tasks. With the growing enthusiasm for AI among technology companies like Apple, Google, Amazon, etc., and the hardware vendors like Intel and Nvidia, the capability of AI assistants is improving each month. These companies keep pushing the state of the art of AI technologies. The book, "Augmented: Life in the Smart Lane," paints an extraordinary vision of how the AI assistants and the cloud of other learning-based and robotic devices will be central players in our human lives in the near future. How well the accuracy of these AI devices catches up with the hype and impacts their enthusiastic followers is something that we would want to monitor as AI innovations become widespread.

There is a variety of AI assistants available on the market that perform various tasks ranging from informing and helping in decision making to helping with other tasks like reading, music playing, event scheduling, and more. These AI assistants learn about the user during usage and adapt to the user, providing personalized services. Some popular AI assistants in the market include Apple's Siri, Amazon's Alexa, Google Now, Microsoft's Cortana, and Samsung's Bixby. Google Assistant is an advanced assistant that combines the power of Google Search, Maps, Play, and YouTube, Mastercard, Allstate, and other companies.

4.2. Integration with Smart Devices

The easyHome work shows an easy way to integrate any smart plug with a third-party system that presents remote management capabilities, with an intuitive user interface. With our developed proof of concept, users can easily integrate any smart plug by simply informing the IP address/hyperlink to the easyHome web page.

Nowadays, there are many industries related to the concept of the Internet of Things. Every year, more startups appear, launching gadgets to the market. Many of these products are created as peripherals that allow controlling traditional electrical appliances. These gadgets are designed to be integrated with big platforms or firms like Google, Alexa, Samsung, Apple, etc. These companies need to create partnerships with third-party firms, or they need to add different adapters so the devices-to-be-integrated can be interconnected.

During the last year, the stand-alone smart devices market has increased rapidly. These devices, such as WiFi smart plugs, are very easy to install and have low prices, so users can acquire a great variety of plugins. In this work, we present easyHome, a system that allows integrating any smart plug with any third-party system that presents remote management capabilities, without adding cost to platforms or plugins. Instead of developing adapters, our flexible link could easily be integrated with future technologies, gadgets, devices, platforms, and systems.

Growing interest in smart home systems reflects the desire of families to manage and control home electrical appliances. This management and control act as an intelligent way to save energy and make life and homes more intelligent, secure, efficient, and modern. Currently, specific firms or platforms develop these kinds of systems, and users need to buy their specific supported devices. This has caused severe user costs and decreased the rate of acceptance of smart systems.

AI Gadgets in Health and Wellness

A high-impact AI-powered neurosimulation gadget called Vertisense is introduced for monitoring many health parameters in real time. To meet the challenges of IoT solutions, in the first author's High Tech Wireless Dynamics knowledge center, his doctoral student has developed a High-Tech Ozone Free-Receiver on-a-Chip to obtain PM2.5, CO, NOx, CO2, and Ozone concentrations using Tin Dioxide Nanorod Sensors for Smartphone Domestic and Wearable Devices. More health-related AI gadgets will be solely discussed in the following paragraphs.

5. AI gadgets in health and wellness. The next couple of sections will focus on AI gadgets for health and wellness. While the majority of AI gadgets are meant for low-level routine tasks, high-level AI tasks require a new generation of robots and AI-powered devices, aiming to replace humans for many tasks. This is either because some tasks are too dangerous, or there are not enough people who want to do certain work, or it's just absolutely impossible to keep enough people on hand for some future repetitive jobs that pay little or are considered too disagreeable for humans to handle. Increasingly, AI robots and gadgets are going beyond routine chores by transforming wellbeing and healthcare delivery, particularly given the aging population and soaring healthcare expenditures.

5.1. Monitoring and Tracking Features

However, when it comes to matters of privacy, this is an area of worry. Many different products of AI gadgets are spinning up these gadgets as cozy and always-present personal assistants, and they train users to share every detail of their day-to-day life. These companies are promising that the more people use their features, services, and devices, the more improvements can be made over time. However, increasing the use of them means losing personal information, which can also put people's privacy at risk. These gadgets require a brief conversation about some of the ways in which some reputable companies limit the risks to privacy. The gadget can also monitor the sleep of the user and the user's partner. However, users can restrict this monitoring to their physical activity only. Users can get data like the amount of sleep, time awake, and body temperature. They can also measure changes in room and environmental temperature, which can include humidity and ambient light for the user. The gadgets can wear the user using the wristband continuously. This means that the device can also track the activities and the fluctuations in body temperature. These monitors also provide beneficial health benefits when they are worn by the user.

Monitoring and tracking features are among the most impressive functions produced by AI gadgets. By collecting data from the things they monitor, they can provide users with powerful insights. These AI gadgets use trained algorithms stored in the cloud. AI gadgets now run many tasks off personal mobile devices and some partially from the edge. There are devices that people are currently using, which can monitor people's body temperature, respiratory rates, and heart rate as well. Drones are another product of AI gadgets that people use in their daily lives, which is a physical form of AI, and it has some built-in intelligence. The gadgets that can navigate themselves around a room are already advanced products of the different fields of computer science. Developers have also advanced these things and are currently using them. These gadgets can map out a room and recognize individual objects.

5.2. Impact on Fitness and Wellbeing

Smart AI-driven fitness gadget design is advisable to increase fitness improvements. Other features improve user acceptance, for which wearable gadgets score very well. Smart wearable AI gadgets incorporate fitness monitoring to track parameters such as distance traveled, calories burned, heart rate, and other physical movements. It's possible to combine both information and communication functions, improving customer acceptance of these AI gadgets. There are many types of fitness designed gadgets that are currently available, including fitness bands, fitness rings, footpods, and AI earbuds. The future of these novel solutions in fitness and well-being is expected to go far beyond embracing these current options, especially with the introduction of AI technology, incorporating machine learning and data analytics. Fitness equipment includes large-scale gadgets – some of which are commercially available today, such as smart cardio machines and smart mirrors that provide a complete workout solution. These advanced sensors capture bodily and facial recognition details, which are processed by AI algorithms. The centralized unit is able to incorporate body and behavior statistics and provide feedback, thus enabling the use of fitness gadgets.

Today, our fitness is a priority because we have a plethora of applications and gadgets that can help us monitor our fitness levels. Telemedicine and virtual checkups can keep us healthy and fit without human intervention. AI-based fitness assistance and AI robots in hospitals and clinics give top-notch care to ensure our well-being and safety. They can visually assess your fitness levels by measuring your calorie intake and suggesting a balanced diet. AI algorithms keep track of what you are eating and drinking with the help of sensors available in smart gadgets. They monitor your daily activities and give you a better understanding of your fitness levels by suggesting proper workouts and giving feedback on different medical and fitness-related inquiries. AI-driven electronic devices can help you monitor your posture if it is bad and suggest improvements based on AI modeling. They ensure you practice self-care by providing alerts or reminders for important medication or health checkups.

AI Gadgets in Home Automation

Rapid advances in technology are transforming the way we live, manage households, and interact with others. The internet of things significantly speeds up this technological makeover via digitally controlled devices that can simplify and improve everyday living situations. The coming generations of domestic IoT devices are also getting smarter, changing their individual behaviors in accordance with their owners' situation and preferences. Smart home technology will be life changing for many people. From wearable devices to apps that track personal fitness and home automation systems that can be operated from any smart device, the way we connect with our surroundings is evolving. Eventually creating an environment distinctly different from the one we know today. More sensitive robots or automation systems fitted with artificial intelligence learn to understand their owner's voices and routines, predict related necessities, and tailor their actions according to the user's situation and requirement.

Home automation means automating the process of controlling home electronics to enable them to work together. IoT, a rapidly growing sector that envelopes both industrial and everyday consumer devices, leverages this. The mainstream adoption of smart home devices has led to remarkable progress in IoT as well. Using wireless communication technology, IoT connects smart devices to the Internet cloud, allowing users' devices to act as a major part of daily life. The dream of being surrounded by intelligent things that are constantly at our service is no longer distant. Most smart home devices, also referred to as the Internet of Things (IoT), are equipped with lots of advanced sensors that can gather and convert dynamic or static signals from the living environment.

6.1. Smart Home Ecosystems

6.1.1. Differing Views of Smartness The smart home is part of a larger conversation about the interaction of people with the space they inhabit. The idea of the smart component has different connotations, depending on industry focus. In the IT world, smart components are devices that have integrated sensors and microprocessors to measure things, make decisions, and take action to improve life. As computing capabilities have extended in recent years, then small, resource-constrained devices have been able to perform advanced functionality at the edge of both the network and the internet. These smart devices have become important players of distributed hybrid IT infrastructures. In the building world, much of the talk of smart cities and smart buildings has centered on making the built environment responsive and reactive to the high-use demands for energy output and utilitarian services such as lighting and ventilation. In the consumer electronics space, the term smart home has, of necessity, encompassed all the other meaning levels as well as its unique set of networking and service labilities.

The smart home is an organism. Each region within it - garage, kitchen, bathroom, bedroom, living space, and returns - is another part of its greater whole. Everything communicates and collaborates through the digital infrastructure which enables it to act autonomously or on command. And the person central to all this networking, data processing, and decision-making is the homeowner. They are the main operator of the environment, communicating their preferences to it through voice, touch, and on-screen menus, as well as by their behavior and previous choices. In turn, the smart home learns and adjusts to its residents' wishes, often trying out innovative solutions on them.

6.2. Energy Efficiency and Sustainability Benefits

Solving the energy efficiency conundrum will require hardware and software collaboration. From the hardware side, a radical change seems in order. Many AI gadgets are aimed at position-related sensing, detection, and robot-like performance. VR goggles are an intermediate valance point because of the need for high-resolution lightweight displays. There is a need for processing ability, but the tremendous clock speed of a top-notch mobile processor will only produce increasing overheating. A more modest, cooler processor will suffice and could even work for less than real-time operations. The shorter processor loading permits lighter phone construction and a larger effective battery size. The VR solution is essentially the current phone model with a new display. AI goggles could work similarly using either chip in a smaller package, more battery, or both. Current standalone VR sets show the power of this approach; they only host the display and assorted controls with the power generator remotely located.

There is also an associated trade-off with reduced charging and battery use. Charging batteries is an energy-intensive process. Most handheld AI gadgets today rely on standard lithium-ion battery technology. Their batteries are typically charged to 80% of maximum charge in 30 minutes. This fast charge comes at a cost: lithium ions are forced into the battery fast as the law of physics allows, producing lithium metal deposits on the anode surface of the battery. After repeated charging, a high surface area of lithium metal reduces battery life. Heat is a problem. AI gadgets are minicomputers carefully designed to dissipate heat, but the more rapid the charging, the more potential remains for damage to internal circuits. Rapid charging is already an energy-saving feature. Suppose you could charge your phone in 5 minutes for 10% depletion. You charge it again 5 times for the 50% used. In the alternative reality where each smaller charge is charged slowly to completion, you would consume three times as much energy.

Ethical and Privacy Considerations in AI Gadgets

The drivers of ethics in AI gadgets are user-centered design, quality of life, and responsibilization; those promoting ethics are values awareness and mindfulness. The Sentient System project proponents proposed talk-based systems as a more user-friendly, flexible, and collaborative interface, offering explicit user consent in joining public home settings to community activities. They also recommended talk-gesture as the next frontier in AI gadgets. Users proposed prankster detection capabilities in AI emotional sensing technology. Ethical AI is also tied to usability, usefulness, and voluntary informed consent in technology products. In addition, AI ethics for gadgets is tied to security concerns, making AI product innovation too expensive to guarantee user safety, particularly during fallible prototype testing. Ethically driven technology should make the internal and external aspects of AI perceivable and comprehensible. Ultimately, AI ethics for gadgets raises user acceptability, and through its business competences, public acceptance of AI gadgets was seen as a form of urban trustworthiness.

Although not typically thought of as "ethical," AI and smart gadgets challenge users' and urban dwellers' values. Privacy concerns were cited across all the gadgets and designs. Wearable trackers and smart home assistants can detect user location and deliver targeted listening or directed advertisements. These marketing practices challenge the integrity, trust, and competence of companies, causing users to alter their comfort and domestic situation. Unsolicited data collection was described as not only a potential violation of the user, but also of other inhabitants within the domestic environment, damaging social and trust relationships. The eavesdropping ability of AI gadgets and being always-on could prevent future useful functionalities or hinder adoption. Privacy, its use, and misuse bleach social values within environments of trust and leisure, redefining homes and relationships away from public consumption.

7.1. Data Security and Privacy Policies

The design and deployment of numerous favorable AI gadgets are accompanied with open challenges and potential risks, such as training on dangerous or offensive content, appropriateness of human-to-AI interactions, mismatching user expectations, data security, privacy, and personal data safety. Comprehensive investigations of the potential negative consequences are provided, and practical precautionary measures for general users are recommended with attention to children, the elderly, and adults with special needs. These considerations serve as valuable discussions among AI researchers, engineers, ethicists, policymakers, and healthcare professionals, especially when the AI models are trained on inappropriate, unethical, or biased data, or deployed freely accessible AI technologies without human supervision. Overall, under concerns of the potential safety risks and the numerous benefits that AI gadgets may bring us, it is critical for AI researchers and industry to work together to propose and continuously improve effective AI safety standards in future guidelines and regulations.

Machine learning and deep reasoning techniques are pioneered for significantly advancing artificial intelligence. Meanwhile, growing data from human interactions and online transactions inspire further applications with compelling user experiences, such as chatbots and lifestyle recommendation systems. With the solid AI algorithm support, a wide spectrum of AI gadgets, such as smart speakers, interactive home robots, and intelligent emoting avatars, are consequently developed from the technology industry. Consequently, AI gadgets have already transformed our everyday lives by providing versatile and natural user interfaces including voice commands, computer vision, and natural language processing. In the near future, AI gadgets have great potential to significantly increase accessibility of such professionally designed AI and technologies for communicating with machines and overall AI experiences, especially for children, the elderly, and adults, to enrich human lifestyle and well-being.

7.2. Ethical Use of AI Technology

On the more appreciative hand, AI cameras and monitors can play a significant role in the protection of product and worker on an industrial base or in hospitals and airports, quickly address murder and theft investigations, and document incidents at home.

While many possibilities have been pursued, AI-related topics and challenges have not been a priority in R&D activities on web-based solutions for emergency response. AI technology could be implemented on gadgets without human-like company information in marketing and advertisements but there could be much more about this area. It is up to individuals mainly to determine the use of tools and whether the benefits of AI are already derived. Even when aware lag, numerous repercussions could flow from a person, organization, industry, or nation.

In this regard, we use our expert interviews with Prof. Veit Etzold, operated labour protection of the COBOT competence center in the way4cs H100 project. As an emergency responder expert, the fourth MCX project interview partner questions whether research has dealt adequately with the potential that AI gadgets are with former civil R&D research projects, participating forces in tends to make emergency operations smarter and safer.

These watchful guardians imply that AI can be used to safeguard and protect people from risks or danger. The ethical use of AI gadgets, for example, wearable cameras that document the work of police officers, can lead to a reduction in crime rates, secured borders, and a higher level of safety.

Future Trends and Innovations in AI Gadgets

The advance of AI gadgets is allowing the transfer of analogous tasks pursued through the smartphone to another series of markets. On the smartphone, it is already possible for the user to ask about the weather, traffic, carry out computer tasks, in summary, make homely life through a gadget. However, there are multiple tasks, such as answering a telephone the moment it is lost, that should be fulfilled or should be better completed through a device other than a mobile. This has led to the appearance of AI gadgets in markets such as sleep management, interaction with potential consumers, and many others.

Artificial intelligence might have come to the forefront with the term AI, but the reality is that AI has been embedded in the technology industry for decades. The overall industry is already used to dealing with AI on a daily basis, but the average user has a hard time leveraging it other than through applications like automatic photo and video editing in smartphones or virtual assistant services. The entry with both feet in everyday life of artificial intelligence came about with the advent of gadgets that offer intelligent functions, where voice assistants stand out and which have been increasing in numbers, functions, and performance.

8.1. Emerging Technologies and Prototypes

8.1.2. Prototype of a Smart Fridge Too many refrigerators contain spoiled food, even though it is hard to finish off all the leftovers quickly. Modern technology can analyze a consistent range of recipes and suggest menus that use the ingredients in a refrigerator. This prototype of a smart refrigerator, developed by a graduate student at the University of Tokyo, has an AI recipe generation function that can analyze various web-based recipe databases to suggest a combination of dishes that share certain common ingredients. It integrates a tablet computer to generate user interfaces. While selecting the dishes, the pantry status is graphically presented. The prototype refrigerator was launched with a web-based recipe database and some pre-defined tools. It was developed for a Chinese target group, but the database can be extended to other countries.

8.1.1. Latest AI Gadgets A wide variety of AI gadgets are being developed worldwide. These products have a common purpose: to bring smarter and faster solutions to everyday life. Currently, there are AI washing machines, dishwashers, air purifiers, intelligent microwaves, and rice cookers available on the market that redefine the original appliance functions. With the rapid development of AIoT technology, connecting products and providing customized services is already a focus of many AIoT brands. AI toothbrushes, remote light switch controllers, Wi-Fi coffee machines, and AI-based beauty mirrors feature a remarkable use of AI technology and are entering more and more How Tech/Smart Tech exhibitions. Some of the latest AI gadgets are examined below.

8.2. Potential Applications in Various Industries

Arguably, the healthcare sector has the most to gain from AI gadgets, particularly in the monitoring and diagnostics of diseases. From small gadgets that can detect anemia to devices that constantly monitor vital signs such as the pulse, respiration rate, and oxygen saturation, to more complex gadgets that measure cancer biomarkers or screen for sexually transmitted diseases, there is great potential for AI gadgets to help us stay healthy. Such devices could coordinate with each other using IoT technologies and send updates to healthcare professionals, making telemedicine safer, more effective, and efficient. For example, wireless sensors with AI capabilities can predict the deterioration of a patient's condition before any signs or symptoms become apparent.

In this final section, we discuss a few examples of how AI gadgets may be utilized in various industries. The examples presented are not exhaustive and are meant to drive home how AI gadgets can be used to enhance different aspects of human life. There are some sectors, such as the agricultural sector, that we did not cover but which will be significantly transformed by AI gadgets, particularly in conjunction with other Industry 4.0 technologies, such as IoT and robotics.

Milton Keynes UK
Ingram Content Group UK Ltd.
UKHW040740301124
451843UK00010B/229